LIGHTING UP THE SKY

ECLIPSES

SETH KINGSTON

PowerKiDS press
New York

Published in 2021 by The Rosen Publishing Group, Inc.
29 East 21st Street, New York, NY 10010

First Edition

Editor: Elizabeth Krajnik
Book Design: Reann Nye

Photo Credits: Cover muratart/Shutterstock.com; series art Zoteva/Shutterstock.com; series art Abstractor/Shutterstock.com; p. 5 Matt Anderson Photography/Moment/Getty Images; p. 6 MPI/Archive Photos/Getty Images; p. 7 Suppakij1017/Shutterstock.com; pp. 9, 13 (top) Andramin/Shutterstock.com; pp. 11, 22 IgorZh/Shutterstock.com; p. 13 (bottom) SP rabbito/Shutterstock.com; p. 14 Jonkman Photography/Moment/Getty Images; p. 15 charlzalan/Shutterstock.com; p. 17 Dudarev Mikhail/Shutterstock.com; pp. 19, 21 Courtsey of NASA; p. 20 ssucsy/E+/Getty Images.

Cataloging-in-Publication Data

Names: Kingston, Seth.
Title: Eclipses / Seth Kingston.
Description: New York : PowerKids Press, 2021. | Series: Lighting up the sky | Includes glossary and index.
Identifiers: ISBN 9781725318410 (pbk.) | ISBN 9781725318434 (library bound) | ISBN 9781725318427 (6 pack)
Subjects: LCSH: Eclipses–Juvenile literature.
Classification: LCC QB175.K56 2021 | DDC 523.9'9–dc23

Manufactured in the United States of America

CONTENTS

LOOK UP!

Something has moved in front of the sun and the skies are darkening. You're wearing special glasses. You're ready for the **solar** eclipse! An eclipse is an **astronomical** event that occurs when a **celestial** body is completely or **partially** obscured, or covered up, by another. Planets orbit around the sun and moons orbit around planets. When three celestial bodies line up during their orbits, an eclipse occurs.

On Earth, we experience solar and **lunar** eclipses. The type of eclipse depends on the order in which the celestial bodies are lined up. Read on to learn more about eclipses.

A transit occurs when a small celestial body eclipses a larger celestial body. An occultation occurs when a celestial body that appears to be large, such as the moon, eclipses a celestial body that appears to be small, such as a star, making the small celestial body disappear.

ECLIPSES IN HISTORY

Before modern science could explain how and why eclipses occur, people throughout the world had different beliefs about what eclipses meant. To some people, an eclipse meant something bad was going to happen, such as war or the end of the world. Other people thought eclipses were caused by dragons swallowing the sun or moon.

SEEING THE LIGHT

In 1919, studying an eclipse helped Albert Einstein prove his **general theory of relativity**, which explains how objects in space move.

Greek **astronomer** Anaxagoras, who lived during the fifth century BC, discovered that the moon and sun are rocky celestial bodies, not gods. He used these discoveries to explain what causes eclipses.

On March 3, 1223 BC, in Ugarit, Syria, historians marked one of the earliest recorded total solar eclipses on a piece of clay, which was discovered in 1948. Around 250 BC, an eclipse convinced Aristotle that Earth is round. In more recent times, eclipses have helped scientists make important discoveries.

SOLAR ECLIPSES

A solar eclipse occurs when the moon comes between Earth and the sun, casting a shadow on Earth's surface. The three types of solar eclipse are total, partial, and annular.

During a total solar eclipse, the sun is completely covered by the moon, the sky becomes very dark, and the temperature drops. This moment is called totality. During a partial solar eclipse, only part of the sun is covered by the moon.

An annular eclipse occurs when the moon is at its farthest from Earth and the sun is nearest to Earth. During this time, you can see the corona, or outermost part of the sun's **atmosphere**.

SEEING THE LIGHT

Even though the sun's light is blocked during a solar eclipse, it's important not to look directly at the sun unless you're wearing special glasses. The light from the sun is so powerful that it can hurt your eyes.

SOLAR ECLIPSE

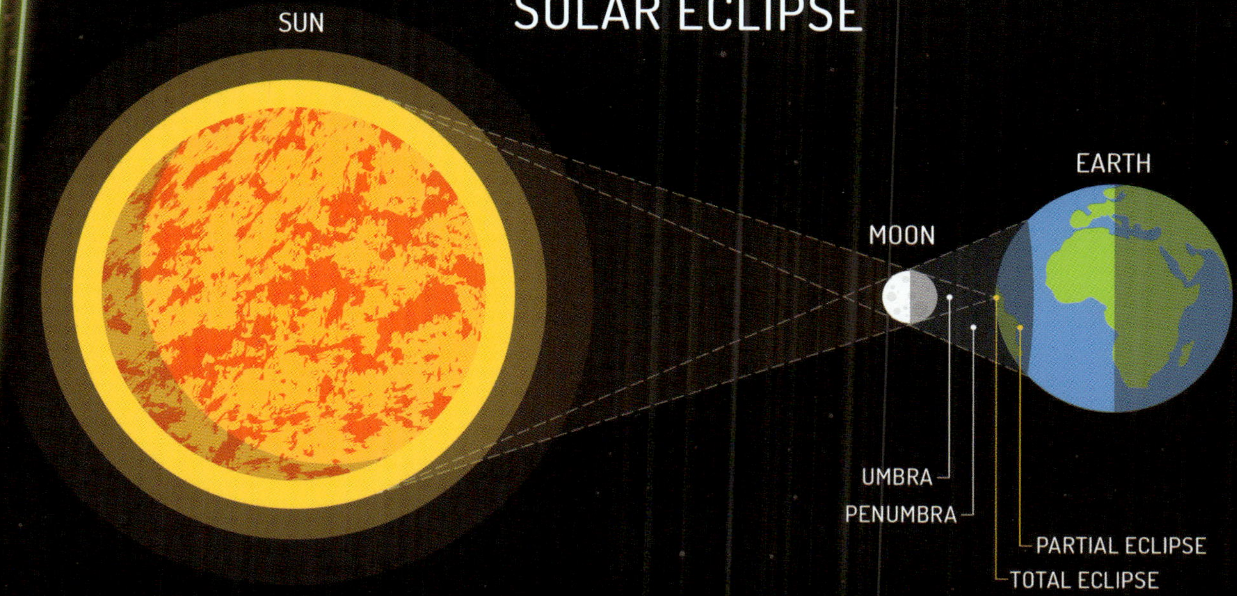

The moon's shadow has two parts: the umbra and penumbra. The umbra is the part of the shadow in which no sunlight can be seen. The penumbra is the part of the shadow in which only some sunlight can be seen.

SOLAR ECLIPSE TYPES

FREQUENCY OF SOLAR ECLIPSES

It takes about one month for the moon to completely orbit Earth. A solar eclipse only occurs during a new moon. However, a solar eclipse doesn't occur during every new moon. Some months, when the moon is closest to the sun, its shadow falls above Earth. Other months, when the moon is farthest from the sun, its shadow falls too low.

The moon's orbit is tilted about 5 degrees toward the **ecliptic**. A solar eclipse only occurs when the new moon is near the ecliptic at either the ascending lunar **node** or the descending lunar node. Most years, Earth experiences between two and five solar eclipses.

SEEING THE LIGHT

How long a solar eclipse lasts depends on what type of eclipse it is and other factors. A total eclipse usually lasts a couple of hours, with totality only lasting a couple of minutes.

On average, Earth experiences a total solar eclipse every 18 months. Only observers standing in the umbra will see a total solar eclipse. The umbra moves across Earth's surface, creating the path of totality.

LUNAR ECLIPSES

A lunar eclipse occurs when Earth comes between the sun and the moon, casting a shadow on the moon's surface. The three types of lunar eclipse are total, partial, and penumbral.

During a total lunar eclipse, Earth blocks most of the sunlight that reflects off of the moon, causing the moon to become dim. However, you can still see the moon. This is the moon's period of totality.

A partial lunar eclipse occurs when only part of the moon passes through Earth's umbral shadow. The moon goes through partial lunar eclipses before it reaches totality. A penumbral lunar eclipse occurs when the moon passes through Earth's faint penumbral shadow.

SEEING THE LIGHT

If you were an astronaut standing on the near side of the moon during a lunar eclipse, you'd actually be seeing a solar eclipse. Earth's nightside would be in front of the sun and you'd see the corona.

The shadows Earth casts onto the moon are also called the umbra and penumbra. Unlike solar eclipses, it's safe to look at lunar eclipses.

BLOOD MOON

Total lunar eclipses are often called blood moons because they look reddish brown or copper during totality. The moon takes on this color because the sunlight that reaches the moon passes through Earth's atmosphere, which scatters the white light in different directions.

SEEING THE LIGHT

The brightness of a total lunar eclipse depends on the conditions of Earth's atmosphere. The eclipse's brightness is given a number from 0 to 4 at the time of mid-totality. This is called the Danjon scale.

At the very beginning and the very end of totality, you might see a blue band on the moon. This occurs because part of Earth's atmosphere called the ozone layer scatters red light and lets some violet and blue light pass through to the moon.

Colors that have shorter **wavelengths** and higher **frequencies**, such as violet, scatter more quickly and never reach the surface of the moon. Colors with longer wavelengths and shorter frequences, such as red and orange, are refracted, or bent, around Earth. They hit the surface of the moon and bounce back toward Earth, which makes the moon look red.

FREQUENCY OF LUNAR ECLIPSES

A lunar eclipse only occurs during a full moon. However, a lunar eclipse doesn't occur during every full moon. Just like the moon doesn't always perfectly line up with the sun to create a total solar eclipse, Earth doesn't always perfectly line up with the sun to create a total lunar eclipse.

From start to finish, a total lunar eclipse can last more than five hours. Each year, Earth usually experiences at least two partial lunar eclipses. However, total lunar eclipses are rarer. On average, Earth experiences two total lunar eclipses in a three-year period. Every 200 years, Earth experiences three total lunar eclipses in the same year.

SEEING THE LIGHT

As long as the sky is clear during a lunar eclipse, everyone on Earth's nightside will be able to see it.

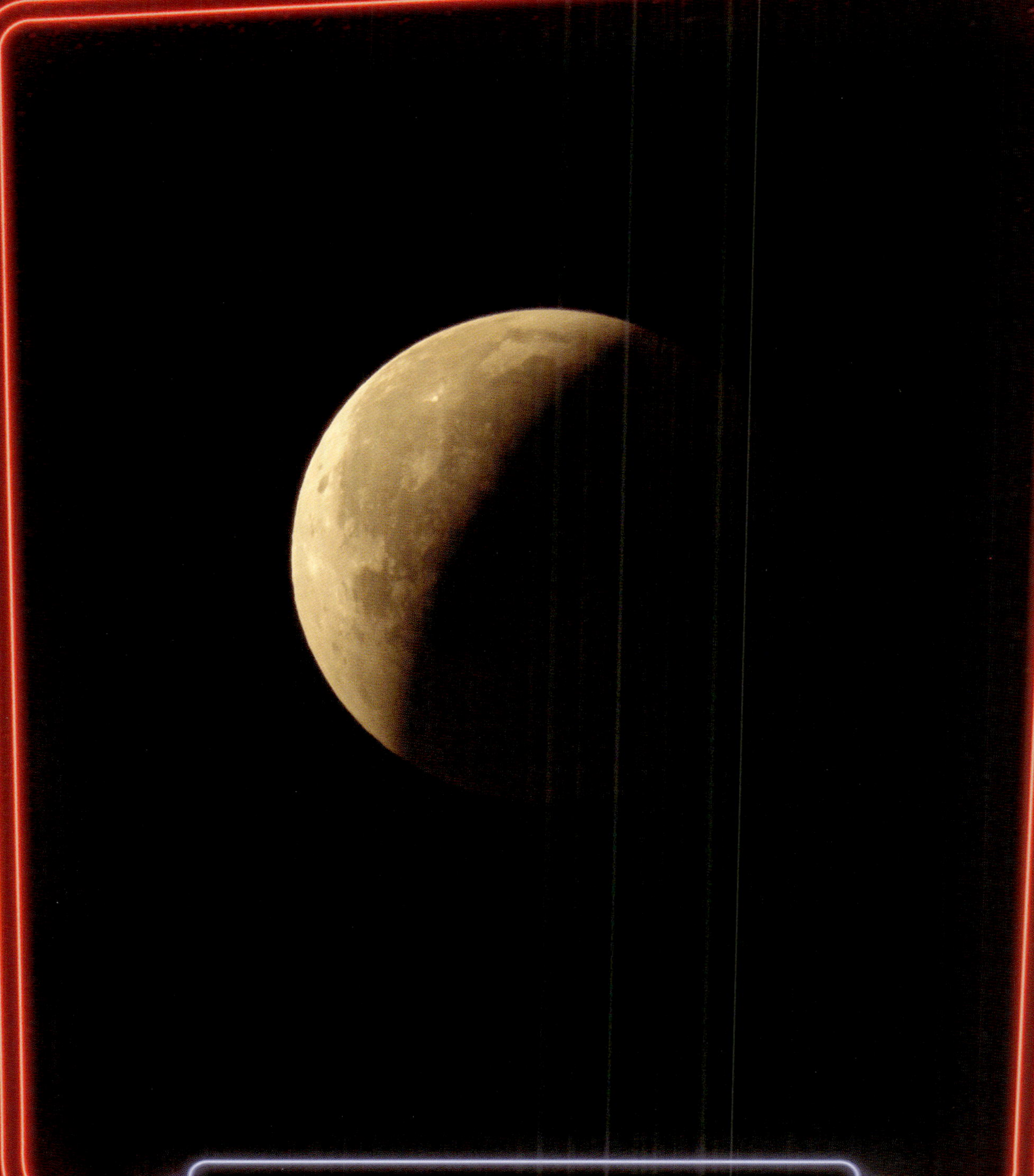

You don't need any special tools to watch a lunar eclipse. However, a telescope will make the moon's surface easier to see.

ECLIPSES ON JUPITER

Earth isn't the only planet that experiences eclipses. Jupiter's four largest moons—Callisto, Europa, Ganymede, and Io—often cause eclipses. This is because, unlike Earth's moon, Jupiter's moons' orbits aren't tilted. With almost every complete orbit around Jupiter, each of these moons causes an eclipse.

Io is the innermost of Jupiter's moons and causes an eclipse every 42 hours. This eclipse only lasts about two hours. Europa's orbit takes almost exactly twice as long as Io's, and Ganymede's orbit takes almost exactly four times as long. Callisto's orbit doesn't often match up with the other moons' orbits.

NASA's Hubble Space Telescope captured this image of three of Jupiter's largest moons—Europa, Callisto, and Io—casting shadows on the face of the planet. This only occurs once or twice every 10 years.

EVERYTHING CHANGES!

The moon, Earth, and all the other planets in our solar system are moving away from the sun. The moon is moving away from Earth at a rate of about 1.5 inches (3.8 cm) per year. Earth is moving away from the sun at a rate of about 0.6 inch (1.5 cm) each year. This occurs because the sun loses mass as it burns, which affects the orbits of all the space objects around it.

Each year, the sun loses about 4.7 million tons (4.3 million mt) of matter. This weakens the sun's gravitational pull on all space objects.

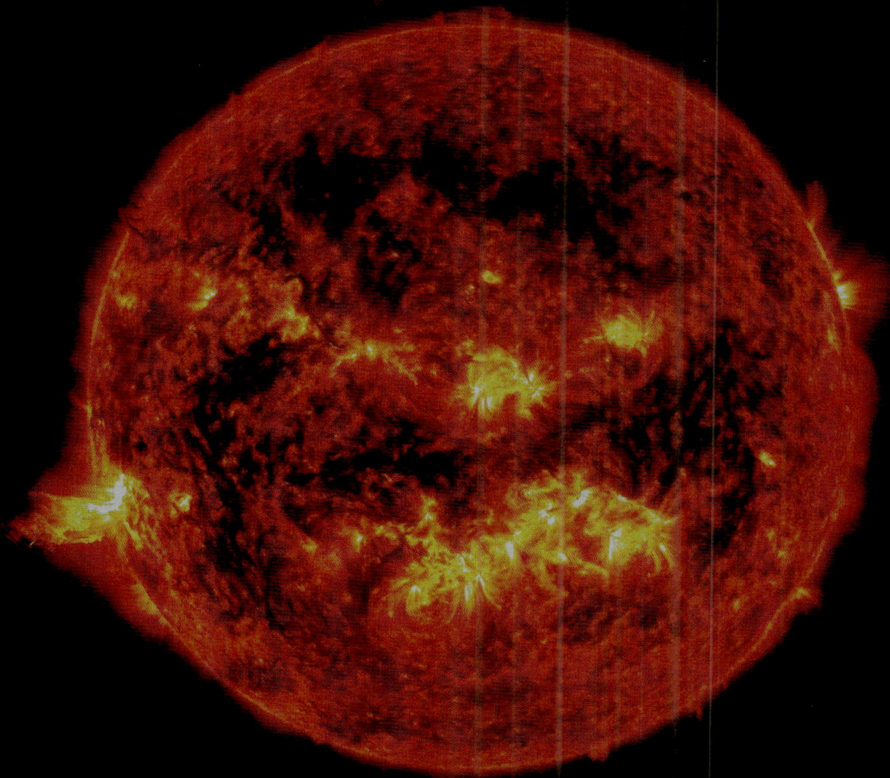

One outcome of this process is that Earth won't experience eclipses the same way millions of years from now. With each passing year, Earth will experience fewer and fewer eclipses.

GET READY TO WATCH

The solar eclipse is about to happen. It's time to get ready:

- CHECK YOUR SPECIAL SOLAR ECLIPSE GLASSES. IF THEY ARE SCRATCHED OR TORN, THROW THEM AWAY AND USE NEW ONES.
- WATCH OTHER PEOPLE. IF THEY DON'T HAVE SPECIAL GLASSES ON, DON'T LET THEM LOOK AT THE SUN.
- IF YOU WEAR GLASSES, KEEP THEM ON AND PUT YOUR SPECIAL SOLAR ECLIPSE GLASSES ON OVER THEM.
- PUT YOUR SOLAR ECLIPSE GLASSES ON BEFORE LOOKING TOWARD THE SUN.
- LOOK AWAY FROM THE SUN BEFORE TAKING OFF YOUR SOLAR ECLIPSE GLASSES.
- YOU MAY ONLY VIEW THE SUN WITHOUT YOUR SOLAR ECLIPSE GLASSES IF YOU ARE INSIDE THE PATH OF TOTALITY AND THE MOON COMPLETELY COVERS THE SUN.
- IF YOU ARE OUTSIDE THE PATH OF TOTALITY, DON'T LOOK DIRECTLY AT THE SUN WITHOUT YOUR SPECIAL SOLAR ECLIPSE GLASSES.

GLOSSARY

astronomer: A person who studies astronomy, which is the science concerned with objects and matter outside Earth's atmosphere and of their motions and makeup.

astronomical: Of or relating to astronomy.

atmosphere: The mass of gases that surround a planet or star.

celestial: Of or relating to the sky.

ecliptic: The great circle of the celestial sphere that is the apparent path of the sun among the stars or Earth as seen from the sun.

frequency: The number of waves of energy that pass by a point every second.

general theory of relativity: Albert Einstein's theory that large objects in space, such as Earth, bend space-time.

lunar: Of or relating to the moon.

node: The intersection of the orbit of a celestial body with the ecliptic.

partially: To some extent, but not completely.

solar: Of or relating to the sun.

wavelength: The distance from one wave of energy to another as it is traveling from one point to another.

INDEX

WEBSITES

Due to the changing nature of Internet links, PowerKids Press has developed an online list of websites related to the subject of this book. This site is updated regularly. Please use this link to access the list:
www.powerkidslinks.com/luts/eclipses